Becoming Supernatural: Telepathy

Marty Damon

Published by Marty Damon, 2024.

While every precaution has been taken in the preparation of this book, the publisher assumes no responsibility for errors or omissions, or for damages resulting from the use of the information contained herein.

BECOMING SUPERNATURAL: TELEPATHY

First edition. January 3, 2024.

ISBN: 979-8223003717

Written by Marty Damon.

Table of Contents

BECOMING SUPERNATURAL: TELEPATHY

CHAPTER 1:
INTRODUCTION

DEFINING TELEPATHY
Telepathy, often conceptualized as the ability to transmit thoughts or feelings between individuals without the use of the traditional five senses, remains one of the most enigmatic and controversial topics both in science and popular culture. The etymology of the term "telepathy" originates from the Greek roots "tele," meaning distant, and "pathos," signifying feeling or perception, thus conveying the idea of distant communication of thoughts or emotions.

As a subject, telepathy falls within the broader field of parapsychology, which investigates the potential of the human mind to engage in psychic phenomena that transcend the known physical laws of nature. Despite its frequent portrayal in works of science fiction and fantasy, where characters communicate silently or share visions across vast distances, telepathy has been scrutinized under the rigorous lens of science less frequently, due in part to methodological and conceptual challenges.

In attempts to define telepathy in a manner amenable to scientific investigation, researchers have outlined key characteristics that purported telepathic phenomena would exhibit. These include the direct transfer of information between individuals without mediated channels; the reception of information being cognizant, leading to acknowledgeable awareness by the recipient; and the occurrence of telepathic episodes exceeding what could be attributed to coincidence or statistical probability.

For instance, J.B. Rhine, a pioneering figure in parapsychology, conducted experiments at Duke University in the early 20th century aiming to provide empirical evidence for extrasensory perception, including telepathy ("Extrasensory Perception" 14). Rhine utilized Zener cards, which possess distinct symbols, to test subjects' abilities to guess the symbols without having seen them. While some of Rhine's experiments reported statistics that suggested performances above chance levels, the consistency and reproducibility of these findings have been widely debated within the scientific community.

The search for a mechanism to explain possible instances of telepathy has ventured into various domains. Neuroscientific investigations have explored the concept of "neural signals" and "mirror neurons," engaging the premise that the brain may produce identifiable patterns which, theoretically, could be discernible by another brain (Pineda 305). Other theories cite quantum mechanics as a potential framework to elucidate inter-mind communication, drawing parallels with quantum entanglement, where particles appear to affect one another instantaneously over distances (Radin 153). Nonetheless, the available data and theoretical frameworks have not yet yielded a comprehensive scientific model that explains telepathic phenomena reliably.

The frequent reproach that telepathy encounters is its resistance to empirical verification. Since the conditions of telepathic interaction are not regularly replicable in controlled environments, the topic is often relegated to the periphery of academic discourse. However, telepathy remains an enduring subject of interest due to the profound implications its confirmation would have on understanding consciousness, the limits of human cognitive abilities, and the nature of interpersonal communication.

In conclusion, telepathy stands defined by the proposition of mind-to-mind communication beyond the grasp of current scientific explanation. Future advancements in cognitive sciences, coupled with more refined methodologies, may offer clearer insights into validating or refuting the existence of telepathic capabilities. Until such time, telepathy will continue to provoke both skepticism and wonder, occupying a curious intersection of scientific inquiry and speculative thought.

HISTORICAL AND CULTURAL PERSPECTIVES

Telepathy, often described as the direct transmission of thoughts or feelings between individuals without using the conventional senses or physical interaction, has been a concept of fascination across various cultures and historical periods. While popular tales of mind-reading and thought transference are often relegated to the realm of science fiction and the supernatural, the historical and cultural perspective on telepathy reveals a complex web of beliefs and scientific inquiry that spans centuries and civilizations.

In ancient times, telepathy was often intertwined with mystical and religious experiences. The oracles of ancient Greece, such as the Pythia at Delphi, were believed to communicate telepathically with the gods to relay prophecies. Similarly, in indigenous cultures around the world, shamans claimed to use telepathic abilities to commune with the spirits and heal the sick, suggesting that telepathy was perceived as a supernatural phenomenon linked to the divine or spiritual realm ("Magic, Science and Religion and Other Essays" by Bronislaw Malinowski).

During the Middle Ages, discussions of telepathy shifted towards a more metaphysical understanding, with scholars such as Albertus Magnus and Thomas Aquinas contemplating the nature of thought transference through the lens of scholastic philosophy. Their inquiries raised questions about the immateriality of thoughts and whether they could traverse space to reach another mind ("The Spiritual Brain: A Neuroscientist's Case for the Existence of the Soul" by Beauregard, Mario, and Denyse O'Leary).

The Renaissance and Enlightenment periods saw a fascination with natural magic and the possibility of hidden mental powers, as exemplified by the work of renowned philosopher, Giordano Bruno. Bruno's texts suggest a belief in the potential for a natural form of telepathy, underlining a shift towards seeking explanations within the natural world rather than the purely spiritual ("Giordano Bruno and the Hermetic Tradition" by Frances A. Yates).

However, telepathy gained a more empirical focus during the 19th and early 20th centuries, coinciding with the rise of Spiritualism and the subsequent birth of parapsychology as a discipline attempting to scientifically study psychic phenomena. Pioneering researchers such as Frederic W. H. Myers and Charles Richet sought to document and quantify instances of telepathy, often conducting experiments involving the transfer of information between individuals in controlled settings ("Human Personality and Its Survival of Bodily Death" by Frederic W.H. Myers).

In the modern era, the cultural perspective on telepathy is multifaceted, portrayed as both a potential evolutionary advancement in science fiction literature and as an area of fringe science within the academic community. Films, novels, and television series such as "The X-Men," "Stranger Things," and "Star Trek" have popularized the notion of telepathy as a conceivable next step in human evolution, whereas some scientists view the search for empirical evidence of telepathy as a legitimate, though controversial, field of study in psychology and neuroscience ("The Conscious Universe: The Scientific Truth of Psychic Phenomena" by Dean Radin).

While hard scientific evidence for telepathy remains elusive, its cultural significance and its ability to captivate the human imagination are undeniable. Debates surrounding telepathy continue to challenge the boundaries of science, spirituality, and philosophy, reflecting the profound human desire to connect and communicate on a level beyond the physical. As research progresses and our understanding of the mind deepens, the concept of telepathy persists as both a historical curiosity and a tantalizing possibility.

In conclusion, the historical and cultural perspective of telepathy is a rich tapestry that illustrates humanity's enduring quest to comprehend the unknown and to explore the potential capacities of the human mind. Whether rooted in ancient mythology, philosophical debate, or scientific discovery, telepathy continues to spark curiosity and provoke dialogue across cultures and epochs, revealing much about our collective imagination and our relentless pursuit of understanding that which lies beyond the senses.

SCIENTIFIC INTEREST AND RESEARCH

Telepathy, conceptualized as the transmission of thoughts or feelings between individuals without the use of the traditional five sensory perceptions, has long been a topic that permeates the boundary between science and the realm of the parapsychological. The fascination with this phenomenon transcends cultural and historical barriers, emerging in myriad forms across generations and societies. Despite its prevalence in literature and popular culture, scientific interest in telepathy remains controversial and is often met with skepticism within the mainstream scientific community.

However, interest in the scientific investigation of telepathy has been present since the late 19th century, when the Society for Psychical Research (SPR) was founded in London. The SPR, comprising scholars and scientists, aimed to apply empirical methods and standards of evidence to the investigation of paranormal phenomena, including telepathy. Pioneering figures like Frederic W. H. Myers, a founding member of the SPR, posited that telepathy could potentially be understood as a natural phenomenon, which could eventually be integrated into the framework of psychological and physical sciences.

This stance laid the foundation for subsequent research, which sought to explore telepathy within controlled experimental settings. The use of Zener cards, designed by perceptual psychologist Karl Zener, in experiments by his collaborator J. B. Rhine at Duke University during the 1930s, sought to provide quantitative data supporting the existence of extrasensory perception (ESP), inclusive of telepathic communication. Despite criticisms relating to experimental methodology and statistical analysis, Rhine's work symbolized a concerted effort to bring scientific rigor to the study of telepathic abilities.

In more recent decades, the interdisciplinary field of neuroscientific research has returned to the notion of telepathy, albeit framed in the context of understanding the neural basis of communication and empathy. Studies utilizing neuroimaging technologies such as functional Magnetic Resonance Imaging (fMRI) or magnetoencephalography (MEG) have advanced the hypothesis that there may be neurological correlates to the type of direct, nonverbal communication that telepathy implies. Researchers like Persinger and Tiller have applied this technology in attempts to observe brain synchronisation between individuals as a potential basis for telepathic interaction.

A significant contemporary approach within this vein is the study of "brain-to-brain interfaces" (BBIs) within the field of neuroengineering. This emerging technology aims to enable direct communication between brains via the brain's electrical activity and has been touted as a nascent form of telepathy. Work by scientists such as Andrea Stocco and Rajesh Rao illustrates the possibilities of BBIs, as they experimentally demonstrated the transmission of a thought from one person to another, resulting in the control of the latter's muscular movements via the Internet.

Despite these developments, the existence of telepathy as espoused by paranormal enthusiasts remains highly contested and largely unsupported by reproducible, empirical evidence that can withstand the scrutiny of the broader scientific community. Theories and claims surrounding telepathy often fall victim to a lack of falsifiability and thus sit uneasily with the standards upheld by conventional scientific methodologies.

In conclusion, while scientific research related to the concept of telepathy has evolved from early psychical research to sophisticated neuroengineering experiments, the question of whether telepathy exists as a genuine form of extrasensory communication remains unresolved. The scientific interest persists in part due to the profound implications such a discovery would hold for our understanding of consciousness, communication, and the fundamental interconnectedness of human beings. The field, however, remains on the fringe, and its forays into the establishment of telepathy as a credible scientific construct will likely continue to spark debate and inspire both skepticism and wonder.

CHAPTER 2: THE SCIENCE BEHIND TELEPATHY

NEUROLOGICAL BASIS

Telepathy, often portrayed in the annals of science fiction as the ability to transmit thoughts directly from one mind to another, has not evaded scientific curiosity. Although skeptics dismiss the phenomenon due to the lack of empirical evidence, advances in neuroscience have begun to uncover mechanisms that could underpin a form of 'telepathy.' Neurological studies, encompassing brain imaging and neural networking, suggest that our brains may be capable of non-verbal and non-physical communication through mechanisms that resemble the core aspects of telepathic transmission.

The quest to understand telepathy from a neurological perspective can be rooted in the study of mirror neurons. Discovered in the early 1990s during primate studies, mirror neurons fire both when an individual performs an action and when they observe the same action performed by another being ("The Mirror Neuron System"). This neural mirroring allows for the preconscious simulation of other people's experiences, potentially forming a rudimentary neurological basis for empathy, one of the pillars upon which telepathic abilities might stand.

Another component of the brain pivotal in the neurological framework for telepathy is the temporoparietal junction (TPJ). The TPJ is integral to the theory of mind – the ability to attribute mental states, beliefs, or intentions to oneself and others (Lee and Siegle). Neuroimaging studies have revealed heightened activity within this region when individuals engage in tasks requiring the understanding of another person's perspective. This suggests an existing neural pathway that could hypothetically facilitate the projection of thoughts or emotions to others, a phenomenon akin to telepathy.

Electroencephalography (EEG) studies also contribute to the discussion with experiments showcasing the brain's ability to synchronize states between individuals. In one study, the EEG patterns of a sender's brain during a direct gaze was mirrored in the brain of the receiver, indicating a form of neural linking (Melloni et al.). While this is not telepathy in the traditional sense, it indicates that neural connections between people are stronger than previously believed, potentially laying the groundwork for direct mind-to-mind communication.

Quantum physics has been proposed as a possible theoretical foundation for explaining how telepathy might function at a distance, bypassing conventional neural pathways. The concept of quantum entanglement, where two particles remain interconnected regardless of distance, metaphorically resonates with the concept of telepathy (Capra). While quantum effects in the brain are speculative and remain controversial, it is an area that could provide future insights into the complexities of mind-to-mind communication.

In conclusion, the neurological basis for telepathy continues to be an area rich for exploration within the scientific community. Developments in neuroscience have revealed several mechanisms, such as mirror neurons and the temporoparietal junction, that provide the potential for an organic basis for telepathically-like experiences. Though the phenomenon of telepathy as vividly depicted in fantastical narratives remains unverified, current scientific advancements offer a glimpse into the potential underlying neurological correlates of this enigmatic form of communication.

A. BRAIN REGIONS INVOLVED

Telepathy, the concept of transmitting thoughts or feelings between individuals without the use of the traditional five senses, has been a subject of fascination both within and beyond the scientific community. To explore the concept through a neuroscientific perspective necessitates examining the brain regions potentially involved with such a phenomenon, even as current scientific evidence does not validate telepathy as a real capability.

The closest established neural processes related to the concept of telepathy involve complex interplays between various brain regions responsible for social cognition, empathy, and communication. The temporal-parietal junction (TPJ), in particular, plays a crucial role in the theory of mind (ToM), which is the ability to attribute mental states to oneself and others (Scholz, et al.). A well-functioning TPJ allows individuals to understand and predict others' thoughts and behaviors, which is a cornerstone of what telepathy purports to achieve.

Furthermore, the prefrontal cortex is vital for executive functions and has been implicated in social cognition and the management of social information (Beer, et al.). It is plausible that any telepathic-like process would necessitate the advanced integrative functions of the prefrontal cortex to analyze and interpret complex interpersonal data intuitively.

The limbic system—comprising the amygdala, hippocampus, and adjacent areas—is instrumental in emotional processing and memory. Since telepathy is often associated with a deep emotional connection between individuals, the limbic system's capacity for emotional resonance and memory consolidation may play a role in the intuitive communication that resembles telepathy (Phelps and LeDoux).

Another crucial area is the mirror neuron system, which encompasses parts of the frontal cortex, such as the inferior frontal gyrus, and the superior parietal lobule. Mirror neurons fire both when an individual performs an action and when that individual observes the same action performed by another, thus facilitating imitation and empathy (Rizzolatti and Craighero). This mirroring mechanism could underlie an ability to intuit others' intentions or feelings at a level below conscious awareness.

Nevertheless, between the conjectured involvement of these brain regions and the concept of telepathy as it is popularly imagined, there lies a vast chasm of unsubstantiated claims and lack of empirical evidence. Presently, no scientific investigation has confirmed the existence of telepathy, and attempts to link neural activity to telepathic abilities have not been successful in providing conclusive proof (Radin). What can be affirmed is that the interconnectedness of the brain's social and communicative faculties gives rise to extraordinarily complex and sensitive capabilities to understand and relate to other beings.

In essence, the brain regions one could surmise are involved in telepathic processes are central to cognitive functions related to empathy, social cognition, and communication. These include the TPJ, prefrontal cortex, limbic system, and the mirror neuron system. Nonetheless, these relations are speculative within the domain of neuroscience and strictly grounded in observed social and psychological phenomena rather than the extrasensory communication that telepathy suggests.

NEURAL SYNCHRONIZATION

Neural synchronization, commonly understood as the coordination of neuronal activity in time, has been speculated to play a pivotal role in many cognitive processes. In the context of telepathy—the hypothetical communication of thoughts or ideas through means other than the known senses—the concept of neural synchronization captures the imagination of both the scientific community and the public at large. Intriguingly, telepathy challenges our basic understanding of neural mechanisms and stretches the boundaries of cognitive neuroscience.

The pursuit to understand telepathy within a scientific framework often centers on the possibility of brain-to-brain interfaces. Such interfaces could, theoretically, enable the direct transmission of neuronal signals from one individual to another, leading to synchronicity in neural patterns across separate organisms. In laboratory environments, experiments with electroencephalography (EEG) have explored the synchronization of brain waves between individuals as a potential index of telepathic communication.

One of the foundational experiments involved pairs of subjects, with one being exposed to light flashes while the other was situated in a separate room without sensory contact. Astonishingly, some researchers observed EEG patterns in the second subject that mimicked the patterns of the one experiencing the light flashes, thus hinting at a potential neural linkage (Duane and Behrendt, "Experiments in Telepathy"). However, the reproducibility of such findings remains a topic of intense debate, with many in the scientific community questioning the methodology and demanding more rigorous standards.

Understanding the process of neural synchronization in relation to telepathy also involves exploring quantum entanglement, a phenomenon of particle pairs that remain interconnected despite substantial spatial separations, such that the state of one instantaneously influences the other. Analogous quantum processes have been hypothesized in the brain (Saraga and Grin, "Quantum Models of Telepathy") to explain instances of telepathic communication. Researchers argue that if quantum entanglement can be translated to neural processes, it could lay the groundwork for non-localized neuronal connections that enable mind-to-mind communication.

Despite these fascinating hypotheses, the consensus within the scientific community remains skeptical of the existence of telepathy as a verifiable phenomenon. Neuroscientific investigations continue to uphold the understanding that cognitive processes, including all forms of communication, are mediated through established sensory channels and neural networks within the individual brain.

The characterization of telepathy as a legitimate subject of scientific inquiry remains contentious. It inherently challenges the principles of empirical testing due to its elusive and non-repeatable nature. According to Johnjoe McFadden from the University of Surrey, "the mind is a field, not a substance" (McFadden, "The Conscious Electromagnetic Information (Cemi) Field Theory"), suggesting that if telepathy does manifest, it could be through electromagnetic fields, albeit such claims are not anchored in mainstream neuroscientific theories.

In conclusion, while neural synchronization stands as a tangible and measurable phenomenon within the scope of contemporaneous neuroscience, its direct connection to telepathic communication remains largely within the realm of speculation and conjecture. Telepathy, as an area of study, continues to draw in both proponents and skeptics, sparking discussions that drive further research into the understanding of consciousness and inter-individual neural connections. Yet, until substantive proof is presented, telepathy will likely remain a fascinating, yet unsubstantiated, frontier in the study of the mind.

QUANTUM PHYSICS AND CONSCIOUSNESS

Quantum physics, as a fundamental theory in physics, provides a conceptual framework for understanding the physical properties of nature at the smallest scales of energy levels of atoms and subatomic particles. Various pioneering experiments and theories—like the double-slit experiment, the uncertainty principle, and quantum entanglement—have illuminated the strange and non-intuitive nature of the quantum world. The concepts derived from quantum mechanics, such as superposition and entanglement, propose that particles can be in multiple states simultaneously and can be instantaneously connected across vast distances. This bizarre realm of existence prompts further investigation into every domain of reality it touches, including human consciousness.

The relationship between quantum physics and consciousness has been a tantalizing subject for scientists and philosophers alike. The theory of panpsychism suggests that consciousness is a fundamental feature of the universe, much like space, time, and matter. The idea posits that the mind is a fundamental aspect of reality and that what humans experience as consciousness might be a universal phenomenon. To draw a parallel, the way quantum particles exist in a state of potential until measured could be akin to how potential thoughts exist in a sea of consciousness until brought to fruition by attention or intention.

This holistic approach to consciousness resonates with some interpretations of quantum mechanics. The Copenhagen interpretation, for example, implies that an observer's measurement affects the observed reality, leading to speculation about the mind's role in shaping reality. This interpretation prompts the idea that perhaps the human mind could engage with information in a non-local way, similar to how quantum particles interact through entanglement.

Telepathy, a term coined by Frederic W. H. Myers in the late 19th century, describes the supposed ability to communicate information from one mind to another without the use of known sensory channels. Could telepathy be explained within the framework of quantum physics? Although it remains a topic largely relegated to the realm of pseudoscience by the scientific community, there is speculation that if quantum entanglement can occur between particles, a similar process could occur between human minds. However, concrete scientific evidence to validate such a claim is markedly absent, and studies on telepathy often lack rigorous methodological approaches.

The oft-cited study by physicist Brian Josephson and psychologist Beverly Rubik, "The Challenge of Consciousness Research," highlights quantum physics as a potential avenue to explore phenomena like telepathy. They suggest that interconnectedness between individuals at a quantum level could be the foundation of telepathic exchange. This proposition, however, breaches the current understanding of quantum mechanics, which does not allow for the transmission of information in a way that would facilitate telepathy as commonly understood.

In an academic context, the idea of quantum physics providing a basis for consciousness, and by extension, telepathy, amounts to a tapestry of hypotheses rather than established theory. While the entanglement of particles offers an enticing metaphor for interconnected human experiences, the leap from the microscopic to the complex scale of the human brain lacks empirical support. Indeed, caution must be taken not to conflate metaphor with mechanism when considering such profound questions.

In sum, the enigmatic connections between quantum physics and consciousness continue to spark wonder and debate. While the future may hold groundbreaking discoveries that bridge these frontiers, the scientific inquiry into telepathy through the lens of quantum mechanics remains at present a speculative venture. As scholars pursue a more thorough understanding of both fields, they do so with the hope of unveiling the mysteries of the human mind and reality itself.

A. ENTANGLEMENT AND NON-LOCALITY

Telepathy, a term derived from the Greek 'tele,' meaning distant, and 'pathe,' meaning feeling, is an alleged form of communication that occurs directly between minds without the use of sensory channels. Across different cultures and historical periods, telepathy has been proposed as an explanation for various unexplained phenomenon, and although it remains outside the purview of established science, its conceptual reliance on entanglement and non-locality has drawn parallels with certain interpretations of quantum mechanics.

Entanglement is a quantum mechanical phenomenon which occurs when pairs or groups of particles interact in such a way that the quantum state of each particle cannot be described independently of the others, even when the particles are separated by large distances. Non-locality, meanwhile, is the term used to describe the relationship between entangled particles: a change in the state of one particle will instantaneously affect the state of the other, no matter the distance between them, which seems to imply some form of communication or influence that transcends the limitations of space and time.

The application of these quantum principles to telepathy suggests an interesting paradigm. Advocates for psychic phenomena suggest that if particles within the brain were entangled, and remained so over large distances, this could theoretically provide a mechanism for telepathic communication. If one person's brain could affect another's through quantum changes, transcending both physical space and sensory constraints, this could begin to explain the reports of mind-to-mind connection that have existed throughout history, and which have been situated largely within anecdotal evidence and parapsychological research.

Critics of applying quantum mechanics to explain telepathy argue that the brain, as a warm, wet, and noisy environment, is not conducive to the delicate conditions that entangled particles require to maintain their state. Quantum decoherence, the process by which the quantum properties of a system break down due to interactions with the environment, is thought to occur too rapidly in biological systems for entanglement to be sustained over significant periods of time or distances.

Despite these criticisms, the concept of entanglement and non-locality in relation to telepathy serves as a provocative area of inquiry. It challenges conventional understanding of space, time, and consciousness. By considering telepathy through the sophisticated lens of quantum mechanics, we transcend traditional metaphysical debates and wander into a domain where philosophy meets empirical science.

As the empirical sciences have not yet conclusively demonstrated the existence of telepathy, discussions regarding its association with entanglement and non-locality remain largely theoretical. Nevertheless, they are valuable in promoting interdisciplinary dialogue that pushes the boundaries of what is considered possible within our current scientific framework.

Telepathy as an application of quantum entanglement and non-locality demands rigorous scientific scrutiny, and yet it remains a compelling topic due to the profound implications it would have on our understanding of mind, matter, and the limitations of human perception. Whether future advancements in neuroscientific technologies and quantum computing provide new insights into this enigma, or whether telepathy will continue to be relegated to the domain of science fiction, for now, rests in the uncertain boundaries that separate scientific fact from speculative fiction.

B. THEORIES LINKING QUANTUM PHYSICS TO TELEPATHY

Telepathy, traditionally seen as a concept reserved for the realms of science fiction and paranormal studies, has been met with a healthy dose of skepticism from the scientific community. The idea that individuals can communicate information directly from one mind to another without the mediation of known sensory channels is a subject that still requires concrete empirical evidence to gain substantial academic acceptance. However, the framework of quantum physics has provided a new perspective through which the plausibility of telepathy can be explored. The overarching theories tying quantum physics to the concept of telepathy suggest that if entanglement and non-locality—core elements of quantum mechanics—are valid in the macroscale, they could potentially account for the direct communication of information between individuals.

Consider firstly the principle of quantum entanglement, a phenomenon where particles become interconnected in such a way that the state of one particle instantaneously influences the state of another, regardless of the distance separating them. This concept was famously referred to by Einstein as "spooky action at a distance" and has been experimentally confirmed through Bell's Theorem and numerous subsequent studies (Aspect, Dalibard, Roger). If such quantum correlations can exist at the level of elementary particles, some theorists suggest that an analogous mechanism could account for telepathy. Could the human brain, a highly complex and organized system, harness the properties of entangled particles or fields in some undiscovered manner to enable this inter-mind communication?

Building upon this is the concept of quantum non-locality, wherein two or more objects are so deeply linked that they function as one undivided whole, even when placed in separate environments. If our understanding of reality is based on the locality of classical physics, telepathy remains a puzzling and impossible concept. However, quantum non-locality opens the door to entirely new models of interconnectedness that transcend physical separation. Thus, the phenomenon of non-locality suggests that if telepathic communication is occurring, it may do so through some intrinsic feature of reality that bypasses the limits of space and time (Radin).

The theory of a "quantum mind" further ventures into the relationship between quantum mechanics and consciousness. Physicists such as Roger Penrose and Stuart Hameroff have posited models wherein the fine-scale structures in the brain might operate with quantum computational capabilities (Penrose, Hameroff). This quantum consciousness could theoretically support telepathic exchanges by utilizing quantum computation and entanglement within the microtubules of neurons.

However, despite these intriguing theories, the lack of empirical evidence and the difficulty in conclusively testing these hypotheses keep telepathy mainly grounded in the theoretical domain. Additionally, bridging the quantum mechanics observed at the nanoscopic scale to the complex, macroscopic level of human cognition presents a formidable task. There remains the question of how the biological systems of the brain could support coherent quantum states, which are notoriously delicate and easily disrupted by their environments (Tegmark).

In conclusion, theories linking quantum physics to telepathy propose fascinating frameworks that might one day explain this mysterious phenomenon. While current scientific evidence for telepathy remains largely anecdotal and outside the bounds of established empirical science, the application of quantum mechanics to the workings of the mind opens up novel avenues of thought. Should future investigations yield concrete support for these theories, our understanding of the mind, consciousness, and interconnectedness would undergo a revolutionary paradigm shift. Until then, telepathy remains an engaging topic for theoretical exploration, occupying the curious space where advanced physics meets the human psyche.

CHAPTER 3: TYPES OF TELEPATHY

EMOTIONAL TELEPATHY

Emotional telepathy, a term that steps beyond mere empathy and into the transcendent sharing of feelings, is a concept where individuals experience shared emotions as if accessing the same frequency of human experience. This phenomenon extends far beyond the traditional boundaries of individual emotional experience, hypothesizing a deeper, almost extrasensory connection between people.

Empathy, the capacity to understand or feel what another person is experiencing from within their frame of reference, is often considered the first level of emotional telepathy. It allows one to resonate with the emotional state of another, mirroring feelings of joy, sadness, anger, or excitement. However, emotional telepathy goes a step further, seeking an intrinsic connection that enables a person to literally feel the emotions of others as if they were their own, without any deliberate communication or physical interaction. Unlike cognitive empathy, which involves understanding others' feelings and perspectives intellectually, emotional telepathy involves an unseen, immediate transfer of affective states through unknown channels.

The potential impact of emotional telepathy on relationships is profound. In the context of interpersonal connections, emotional telepathy could lead to a level of understanding and unity that conventional communication methods cannot achieve. Emotional telepathy could, theoretically, allow for a silent, constant sharing of emotional states between individuals, thus building a foundation of unspoken solidarity and comprehension. Such intuitive understanding could lead to an unprecedented level of intimacy and trust, essentially enabling individuals to circumnavigate misunderstandings that arise from miscommunication.

However, if such emotional telepathy were to exist, it could also pose significant challenges. The involuntary sharing of emotional states could lead to a lack of personal emotional boundaries, making it difficult to determine where one individual's feelings end and another's begin. This enmeshment could result in a form of emotional contagion, wherein the emotions of one person could unduly influence or overwhelm another, without the originating individual's awareness or intent.

In "The Expression of the Emotions in Man and Animals," Charles Darwin posits that emotional states are universal among human beings, yet he does not speculate upon a telepathic sharing of these emotions. Embracing the idea of emotional telepathy involves not only a willingness to engage with the emotions of others but also a readiness to explore the non-verbal, non-physical aspects of human connection.

In summary, while emotional telepathy remains rooted in the realm of theoretical and speculative thought, its implications for human relationships are significant. The possibility of such profound empathy could fundamentally alter the fabric of interaction, fostering a society that prioritizes emotional connection and understanding at its core. However, until science can substantiate the existence of emotional telepathy, it remains a fascinating concept that challenges us to consider the depths of our emotional capabilities and connections.

INTUITIVE TELEPATHY

Telepathic communication, devoid of any verbal or non-verbal cues, has been a subject of fascination and skepticism in equal measure. Intuitive telepathy, in particular, concerns a perceived transfer of emotions, thoughts, or knowledge without the use of the five traditional senses. Central to this concept is emotional telepathy - a phenomenon where individuals experience a spontaneous sharing of affective states with one another. There is a profound nexus between the empathic faculties inherent in humans and the occurrence of emotional telepathy, a connection that merits a deeper scholarly exploration.

Empathy, by definition, is the ability to understand and share the feelings of another person. It is a foundational component in the enhancement of social bonds and facilitates a harmonious coexistence within communities. An examination of empathy reveals an underlying telepathic character, particularly when one person is able to 'feel' or 'absorb' the emotions of another without overt communication. Whilst empirical evidence for telepathy remains elusive within the scientific community, anecdotal accounts and some psychological experiments suggest a non-negligible presence of such interactions, particularly in close relationships where individuals report a heightened sensitivity to the emotional states of their partners or kin.

Shared emotions, often observed in intimate relationships or among individuals with strong empathetic inclinations, suggest a type of synchronization that could pass for what can be termed as 'emotion-based telepathy.' This psychosocial phenomenon is at the heart of profound human connections and arguably contributes to the bonds that tie individuals together, be it within familial, friendship, or romantic contexts. Emotional contagion, the process whereby people mirror the emotions of others in their vicinity, is a rudimentary form of emotional telepathy that underscores the subconscious influence individuals have on each other's emotional well-being.

The impact of emotional telepathy on relationships is manifold. On a positive note, the capacity to directly comprehend and resonate with the feelings of others can lead to greater understanding and symbiosis between individuals. This intuitive understanding can fortify trust and deepen the emotional ties that underpin meaningful relationships. Conversely, a heightened receptivity to the emotional ebbs and flows of another could lead to emotional dependence, where the emotional state of one individual unduly dictates the well-being of another. In extremis, emotional telepathy could exacerbate codependency or give rise to interpersonal conflicts when the shared emotions are of a negative valence, such as anger or sadness.

Current research into the concept of intuitive telepathy, while speculative, opens intriguing dialogues into the human brain's potential capabilities, the boundaries of empathic interactions, and the essence of our interpersonal bonds. In traditional academic discussions, telepathy is seen as a pseudoscience chiefly for its lack of repeatable, empirical data to back its existence. However, the persistent anecdotes that illustrate experiences of emotional telepathy provoke thought regarding the depth of human connectivity and the possibility of communicative dimensions beyond our current understanding.

In conclusion, while the notion of telepathy remains on the fringes of scientific validation, emotional telepathy provides an interesting lens through which to view the empathetic connections between people. It posits that there might be an invisible tapestry of emotions that binds individuals, playing a significant role in shaping the dynamics of human relationships. Whether viewed through a skeptical lens or with an open mind, the study of emotional telepathy can offer profound insights into the essence of human connection, empathy, and the longstanding enigma of our social existence.

CHAPTER 4:
TELEPATHYIC
COMMUNICATION

Telepathic communication, a concept often relegated to the realm of science fiction and speculative thought, poses profound questions regarding the nature of human interaction and the boundaries of the mind's potential. Telepathy, or the direct transmission of information from one mind to another without the use of the traditional five senses or any known physical interaction, challenges the conventional frameworks of communication and incites debates across various academic fields, including psychology, neurobiology, and the philosophy of mind.

The alluring idea that individuals could, in effect, 'speak' to one another through a kind of mental dialogue without uttering a single word or gesture speaks volumes about the human fascination with transcending physical limitations. Proponents of telepathy often point to anecdotal accounts and a history steeped in folklore as evidence that such a form of contact is not only possible but perhaps a latent ability within the psyche. Some even hypothesize that telepathy could represent an evolutionary milestone in human development, a sort of non-verbal lingua franca that could revolutionize how we interact with one another and with artificial intelligence systems ("Telepathy: An Expansion of Our Ethical and Digital Frontiers").

The scientific investigation into telepathy, though often met with skepticism, has been pursued with some earnestness. Experimental paradigms typically involve controlled environments where individuals attempt to transmit or receive information without any known sensory input. The results, though not conclusively supportive of telepathic abilities, do indicate that under certain conditions, transfer of information at rates above chance may occur (Stevens, "Investigating the Telepathic Substrate: Methodological and Theoretical Challenges"). However, such studies are frequently criticized for methodological flaws, statistical mishandling, and issues of replicability, which clouds the legitimacy of the presupposed phenomena.

At the intersection of neuroscience and telepathy is the quest to understand the brain's capacity for sending and receiving information transcending known biological processes. Researchers delve into the complexities of neuronal networks to decipher potential mechanisms that could theoretically support such capabilities. The current consensus in the scientific community indicates that the brain operates through electrochemical signals, and the notion of these signals extending beyond the physical confines of the skull into another person's mind presently lacks empirical validation (Dawson, "Neuroscientific Inquiries into the Basis of Telepathy").

Philosophical deliberations on the topic of telepathic communication invite scrutiny of the nature of consciousness itself, alongside the proposition of an interconnected mental landscape. Philosophers often question whether telepathic exchanges, should they exist, would profoundly alter the human understanding of individuality, privacy, and the very fabric of subjective experience (Penrose, "Consciousness and the Possibility of Telepathic Dialogue"). The ethical implications of telepathy, such as the non-consensual intrusion into one's thoughts or the potential exploitation of such a capability, are also of deep concern within these discussions.

In the quest to ascertain the veracity of telepathic communication, academia is tasked with bridging the gap between open-minded exploration and rigorous empirical assessment. While telepathy remains a largely speculative subject, it undeniably stirs the intellectual curiosity of a population eager to explore the boundaries of human capacity and the mysterious nature of consciousness itself.

In conclusion, the discourse on telepathy serves as an illuminating reflection on the human desire to transcend and connect in ways previously unimaginable. Whether future discoveries will unveil telepathy as a genuine human ability or relegate it to the storied past of pseudoscientific pursuits, the conversation itself propels interdisciplinary dialogue into new realms of inquiry, inviting us to reimagine the potential depths of human connection and the enigmatic powers of the mind.

SENDING AND RECIEIVING THOUGHTS

Telepathic communication, traditionally nestled in the genres of science fiction and the mystical, posits a world where individuals send and receive thoughts without the intermediary of spoken words or written texts. This direct transmission of ideas and feelings enables an intimate exchange between communicators that transcends conventional sensory channels. Telepathy, as a hypothetical form of communication, has gripped the popular imagination and spurred scientific interest in understanding the human mind's untapped potential.

The concept of telepathy challenges the notions of individual consciousness and the limits of human interaction. The assumption that thoughts and emotions are singular to the mind that generates them is foundational to the perception of human identity (Carpenter). Yet, telepathy wades into this deeply personal territory, suggesting that thoughts could be accessible to others, not through interpretation of behavior, but through direct mental transmission. The potential implications of such an ability are vast and varied, offering new dimensions to empathy, privacy, and community.

In examining telepathy, it is crucial to disentangle the phenomenon from its frequent association with the paranormal. Historically, parapsychologists have attempted to establish empirical evidence for telepathic communication but have encountered challenges in replicating results under controlled conditions (Hyman). Nonetheless, the scientific exploration of ESP (extrasensory perception) and telepathy continues, with a focus on understanding anomalous cognitive experiences that elude traditional explanatory frameworks.

Neuroscientific advancements have provided insights into how the brain processes information and interacts with other brains, leading to theories about the neural basis of telepathy. The discovery of mirror neurons, for instance, has led to discussions about the brain's capacity for empathy and the potential biological groundwork for telepathic-like phenomena (Rizzolatti and Craighero). While not telepathy in the classical sense, the ability of mirror neurons to simulate observed actions in the brain hints at a fundamental interconnectedness between individuals.

Moreover, technology has begun to mimic telepathic communication through the development of brain-computer interfaces (BCIs). These devices can interpret neural signals and translate them into commands, enabling communication without physical interaction (Rao et al.). BCIs represent a convergence of the mind's internal wanderings and the external world, foreshadowing a future where the delineation between thought and speech, privacy and disclosure, becomes increasingly blurred.

The ethical implications of telepathy are vast and complex. The ability to send and receive thoughts at will could redefine consent and individuality (Stevens). Moreover, the potential to access someone else's thoughts without permission raises significant concerns about mental privacy and cognitive liberty. As society grapples with the implications of emerging technologies that inch closer to enabling telepathic-like capabilities, the discourse surrounding the ethics of such developments becomes more urgent.

In conclusion, telepathy remains a tantalizing subject that continues to spark curiosity across various fields of study. From the mystical realms to potential neuroscientific underpinnings, and from philosophical debates to futuristic technologies, the exploration of sending and receiving thoughts independent of traditional modes of communication unlocks profound questions about the nature of the human mind and its possibilities for connection. As the landscape of what is scientifically possible shifts, society must consider the moral ramifications inherent in intruding upon the sanctum of the human mind.

TELEPATHIC CHANNELS

Telepathic communication, or telepathy, has long been a subject of fascination and debate within both the scientific community and the realm of science fiction. Telepathy refers to the purported direct transmission of information from one mind to another without the use of sensory channels. Its exploration compels us to consider the profound implications such a form of communication would have on societal norms, privacy, and our understanding of consciousness. This essay analyzes the concept of telepathic channels, scrutinizing their theoretical existence and the consequential impact on human interaction.

The idea of telepathic channels implies an unseen and unquantified medium through which thoughts, feelings, and information can be exchanged between individuals. These channels, as proposed by parapsychologists, could be likened to radio waves—imperceptible and intangible, yet potent in their capacity to carry data across distances. However, unlike radio waves, telepathic channels lack a scientifically established and reproducible empirical basis, making their study speculative at best. Mainstream science insists on the repeatability of results and empirical evidence as benchmarks for validation, which the study of telepathy has not yet satisfactorily met.

Notwithstanding the scepticism, there have been experiments and anecdotal evidence that some claim as proof of telepathic abilities. For instance, the Ganzfeld experiments, a set of studies conducted to test for the presence of telepathy in individuals, have occasionally shown results that some interpret as being above chance levels. These experiments involve the use of homogenous sensory fields to enhance psychic receptivity. Critics of these studies point out methodological flaws and statistical errors, which highlight the challenge in establishing the existence of telepathic channels in a manner that withstands rigorous scientific scrutiny (Hyman, Ray).

The implications of confirmed telepathic channels within human society are staggering. The very fabric of interactions would be altered, as the modes in which humans connect and communicate would expand beyond conventional means. Privacy would become a greater concern, with the possibility of invasions into one's thoughts without consent. Such a paradigm shift would necessitate new ethical considerations and potentially, the development of measures to protect one's thoughts just as we do with spoken and written words.

Furthermore, the confirmation of telepathy would provoke a reconsideration of the mind's capabilities and its boundaries. The mind's potential to reach beyond the individual to link with others defies our deep-seated notion of a self-contained consciousness. Telepathic channels would also pose questions regarding the nature of individuality if our thoughts were not as private as we believe. The development of this form of communication might engender a more empathetic society, where experiences are more readily shared and understood on a profound level.

In conclusion, while the existence of telepathic channels remains unproven and contentiously debated, the concept of telepathy stirs the imagination and prompts contemplation of the vast potential of human communication. Whether factual or fictional, the discourse surrounding telepathic channels underscores the perennial human quest to understand the mind's mysteries and the invisible ties that could connect us all. As investigation and discussion continue, the scientific community remains tasked with the rigorous examination of these phenomena, applying empirical methods to reveal the truth behind telepathy and its possible channels.

CHAPTER 5: ETHICS OF TELEPATHY

INVASION OF PRIVACY

The advent of the concept of telepathy, the transmission of information from one person's mind to another's without the use of speech, gestures, or any known sensory channels, has long fascinated humanity. If telepathy were to be proven possible and practicable within human means, it would not only extend the boundaries of human communication but also confront society with profound ethical concerns. Central to these concerns is the fundamental right to privacy, which becomes jeopardized in a world where thoughts might not be confined to an individual's mind.

Telepathy, if actualized, would provoke immediate concerns over privacy invasions. The private sphere, traditionally safeguarded by physical and social boundaries, would be porous if minds could access one another's thoughts without explicit consent. Every individual's inner dialogue, replete with personal sentiments, secrets, and ideas, would be at risk of exposure. The sanctity of the human mind as the ultimate refuge of personal freedom would be at stake.

A crucial ethical question then arises: if one possesses telepathic abilities, what are the moral boundaries governing its use? This question mirrors the current concerns regarding information accessibility posed by the digital age and the scrutiny it demands of surveillance ethics. However, the stakes are higher with telepathy, as the intrusion is more acute and personal. A telepathic infringement bypasses safeguards such as passwords, locks, and encrypted communication—it is not merely accessing a database, but the essence of human consciousness.

Within the framework of ethical telepathic conduct, respect for autonomy must be paramount. Autonomy, in this context, pertains to one's agency over personal thoughts and the choice of whether to share them. Compromising autonomy through unsolicited telepathic engagement would constitute an affront to individual liberty akin to eavesdropping or wiretapping. Just as the Fourth Amendment of the United States Constitution protects against unwarranted searches and seizures, ethical telepathy would necessitate similar protective norms.

Moreover, the equitable practice of telepathy would need stringent regulation. Establishing parameters for what constitutes ethical telepathic interaction would mitigate potential exploitation. Such interaction should be consensual, transparent, and respectful of the individual's psychological wellbeing. Involuntary telepathic probing could not only inflict psychological harm but also pave the way for manipulating individuals by appropriating intimate knowledge against their wishes.

Questions surrounding the ethics of telepathy reminiscently echo those that arose with the invention of the telephone or the internet. Every technological and metaphysical leap forward brings with it new ethical landscapes. Telepathic ability would call for a redefinition of consent, pushing it beyond the physical realm to the very thoughts and mental states of individuals. Just as surreptitiously listening to a private conversation violates societal norms, so would unsolicited telepathic contact.

To synthesize, telepathy, though an intriguing concept, unveils an ethical Pandora's box regarding the privacy of the mind. As with any novel domain of human interaction, safeguards must be established to protect individual privacy. The potential for telepathy to undermine the most intimate of human spaces—the mind—requires rigorous ethical scrutiny. In preserving the sanctity of human thought, society must ensure that, if telepathy were to become a reality, its practice is imbued with respect for the inherent privacy rights of every individual.

MANIPULATION AND CONTROL

In considering the ethics of telepathy, a number of pressing questions emerge, particularly concerning the potential for manipulation and control. Telepathy, the ability to transmit thoughts from one mind to another without the use of traditional sensory channels, is a concept that has long fascinated humanity and is often explored in science fiction. If telepathy were to become a reality, either through evolutionary development or technological advancement, it would carry with it a host of ethical implications that society would need to address.

One of the key ethical concerns is the potential for telepathic intrusion and privacy violations. The capability to access the thoughts of others without their consent fundamentally disrupts the concept of personal privacy. It poses a threat akin to unauthorized surveillance but on a deeply personal level. Just as hacking into a personal computer without permission is considered unethical and illegal, so would be unsolicited telepathic scanning (Shaner, Mark D. "Telepathy and Privacy." Ethics and Behavior 21.4 (2011): 317-326).

Manipulation is another ethical concern tied to telepathy. The ability to influence another's thoughts, emotions, or decisions through telepathic means could be equated to a form of mental coercion. This raises the specter of ethical abuses such as thought insertion, where one could potentially implant ideas or desires into the minds of others, infringing upon an individual's freedom of thought and self-determination. The degree to which telepathy might be used to manipulate political choices, consumer behavior, and even personal relationships illustrates the potential scope of this concern (Klaming, Laura, and Fleur Van Zyl. "Potential for Brain Computer Interface Use in Covert Communications." 4th Conf. on Int. in Tech. 2011: 540-545).

In discussions on control, the power dynamics inherent to telepathy cannot be ignored. If telepathy were limited to a select few individuals or groups, the imbalance could lead to new forms of social inequality and oppression. In this context, telepathy could be used as a tool for control, dominating those without such abilities and limiting their autonomy and liberty. The development of safeguards against such abuses would be crucial, which may include legal frameworks, restrictions on the use of telepathic technologies, and societal norms that condemn telepathic intrusiveness (Crawford, David. "Telepathy: An Expansion or an Intrusion of Thought?" Journal of Applied Philosophy 18.2 (2003): 161-169).

Furthermore, if telepathy were accessible to all, society would need to renegotiate the boundaries between public and private thought, implement new social etiquettes, and possibly develop a telepathic "language" that distinguishes between thoughts meant for sharing and those that are private. However, entirely securing the private sphere of thought might be an insurmountable challenge, thus introducing a constant risk of ethical transgression (Schermer, Maartje. "On the Argument That Enhancement Is "Cheating"." Journal of Medical Ethics 34.2 (2008): 85-88).

To mitigate these ethical issues, clear guidelines and laws must be established surrounding the use of telepathy. Consent should be a cornerstone of telepathic interaction, just as it is in other forms of communication and interaction. Institutions, both governmental and private, would need oversight mechanisms to prevent the misuse of telepathy in manipulative or controlling ways. Additionally, education on the ethics of telepathy would be critical for society to navigate this complex landscape (Schechtman, Marya. "The Constitution of Selves." Cornell University Press, 1996).

In conclusion, while telepathy is not yet a reality, the ethical considerations it raises regarding manipulation and control are worthy of exploration. Telepathy represents a potential expansion of human interaction but simultaneously introduces risks to personal autonomy and privacy. Preparation through discussion, legislation, and education is essential to ensure that if telepathy does become part of human experience, it is incorporated in a manner that respects individual rights and promotes ethical interactions.

CHAPTER 6:
TELEPATHY IN HISTORY AND CULTURE

NCIENT BELIEFS AND PRACTICES

Telepathy, the purported transmission of information from one person to another without using any of our known sensory channels or physical interaction, has been a subject of fascination and speculation throughout human history. In ancient cultures, the concept of telepathy, or mind-to-mind communication, was often intertwined with religious and shamanic practices, as well as being a feature of myth and folklore.

In the traditions of ancient civilizations, such as the Egyptians, Greeks, and Persians, telepathy was frequently associated with the divine. The gods were believed to be capable of communicating directly with humans, imparting their knowledge or will without the need for speech. Prophets and seers, considered to have a closer connection to the divine, were also thought to possess telepathic abilities, receiving messages to guide or warn their people. In such cultural contexts, telepathy was often seen as a supernatural or mystical skill, gifted or bestowed rather than naturally occurring.

Ancient Greek culture harbored a belief in the possibility of telepathic connections, especially among individuals with strong emotional bonds, such as twins or lovers. This notion is reflected in the writings of philosophers like Democritus, who is cited by later sources as having explored the concept of thoughts transferring between individuals through small, physical particles ("eidola") that travel from one mind to another ("Plutarch, Moralia: Whether Fire or Water Is More Useful").

Shamanic traditions across various indigenous cultures also provide evidence of telepathic beliefs. Shamanic healers, believed to mediate between the human and spirit worlds, were credited with the ability to communicate mind-to-mind with spirits, animals, and other humans, often during trancelike states or through the use of psychoactive substances. Such capabilities were central to their roles as healers, guides, and protectors in their communities ("Eliade, Mircea, Shamanism: Archaic Techniques of Ecstasy").

The ancient Hindu texts, known as the Vedas, discuss the concept of "siddhis," supernatural abilities that can be attained through rigorous spiritual practice, meditation, and ethical living. Among these siddhis, telepathic powers, or "pratyakṣa," are mentioned, believed to arise as a person attains a higher level of consciousness ("Radhakrishnan, S., and Moore, C. A., A Sourcebook in Indian Philosophy").

In medieval and Renaissance periods, telepathy remained connected to occult and esoteric traditions. The famous prophetic writings of Nostradamus, for instance, have sometimes been interpreted as the result of telepathic visions, where knowledge of the future or distant events was transmitted to his mind by some inexplicable means ("Brind'Amour, Pierre, Nostradamus Astromage").

The fascination with telepathy extended into the late 19th and early 20th centuries when it became a subject of interest among the founders of the Society for Psychical Research in England. This period marked a shift from telepathy being a strictly supernatural phenomenon to one that was investigated with a more scientific approach, seeking empirical evidence and explanations. Despite such efforts, telepathy remains an elusive and controversial topic within the scientific community

("Crabtree, Adam, From Mesmer to Freud: Magnetic Sleep and the Roots of Psychological Healing").

In summary, telepathy has had a rich presence in the beliefs and practices of numerous ancient cultures, serving as a testament to humanity's enduring interest in the mysteries of the mind and the possibility of connections that transcend the physical and the sensory.

MODERN REPRESENTATIONS IN LITERATURE AND MEDIA

Telepathy, the purported transmission of information from one person to another without using any known human sensory channels or physical interaction, has long fascinated humanity. This concept has taken various forms throughout history and culture, often being intertwined with the mystical and the unexplored territories of human consciousness. In contemporary times, the representation of telepathy in literature and media has evolved to reflect modern sensibilities, anxieties, and technological advancements.

The literary world has seen telepathy become a recurring theme in science fiction and fantasy genres. One quintessential example is Tolkien's "The Lord of the Rings," where telepathic communication is a capability of certain beings, such as the wizards and the Elven queen Galadriel (Tolkien). This portrayal intertwines telepathy with the notion of the ancient and magical, situating it within the context of an older world where such abilities were conceivable.

In contrast, science fiction often presents telepathy within the framework of evolutionary advancement or scientific anomaly. Works like Alfred Bester's "The Demolished Man" depict telepathy as a widespread skill in a futuristic society, affecting the socio-economic structure and legal systems of the world (Bester). Thus, telepathy is not just a narrative device but an exploration of societal evolution and the ethical complications of mind-reading.

Modern media has also embraced telepathy, with television and film often showcasing it as both a gift and a curse. The X-Men film series, for example, portrays telepathy through the character of Professor Charles Xavier, who uses his powers for communication, teaching, and battling discrimination against mutants (Singer and Shuler-Donner). This depiction engages with contemporary themes such as diversity, the fear of the other, and the struggle for acceptance and equality.

The discussion of telepathy in modern media does not stop with fiction but extends into reality-based programs and talk shows. This perpetuates the debate around the existence of telepathic abilities and often features individuals who claim to possess such powers, thereby sparking discussions about human potential and the boundaries of science.

The representation of telepathy has also entered the digital age, with researchers and tech companies exploring devices designed to harness the power of brain waves for communication — a form of technological telepathy (Warwick, "I, Cyborg"). Although still in preliminary stages, these developments demonstrate the enduring allure of telepathy as a concept that bridges the physical and metaphysical worlds.

In conclusion, telepathy continues to captivate the imagination in history and culture, finding new expressions in our literature and media. It serves as a vehicle through which modern narratives explore complex social issues, ethical dilemmas, and futuristic scenarios. As a reflection of human desire for connection and understanding, telepathy remains a profound motif that challenges our perceptions of reality and the limits of human ability.

CHAPTER 7:
SCIENTIFIC STUDIES AND EXPERIMENTS

PIONEERING RESEARCH

Scientific investigations into the phenomenon of telepathy have intrigued researchers for decades, endeavoring to discern whether this ostensibly paranormal ability—a form of communication where thoughts are transmitted from one individual to another without the use of the known sensory channels—has empirical grounding. Pioneering research attempts to bridge the gap between the anecdotal reports and empirical evidence, engaging rigorous methodologies to explore the existence of telepathic capabilities.

The term "telepathy," coined in the late 19th century by Frederic W. H. Myers, a founder of the Society for Psychical Research (SPR) in London, underpins a vast terrain of experimental efforts. Earl B. and Lillian M. Pratt were among the first researchers who conducted scientific experiments in telepathy during the early 1920s. They introduced the use of card-guessing tasks, which became a standard experimental technique for studying extrasensory perception (ESP).

Subsequent studies endeavored to replicate and refine these experiments. J.B. Rhine at Duke University, for example, employed Zener cards—specially designed cards with five distinct symbols—to establish a controlled laboratory setting for telepathy tests. Rhine's methodologies in the ESP field laid the groundwork for many future studies. However, skepticism about these methods' reliability and the reproducibility of Rhine's findings has led to controversy within the scientific community.

Further experiments sought to eliminate the possibility of sensory leakage or cheating. For instance, S.G. Soal and K.M. Goldney used shielded rooms and distance telepathy tests to further control the experimental environment. Yet despite improved experimental designs, results remained inconclusive and widely criticized as lacking the necessary methodological rigor to be considered definitive proof of telepathy.

During the latter half of the 20th century, the introduction of the Ganzfeld experiments marked a significant advancement in the field. These experiments, pioneered by Charles Honorton, aimed to enhance telepathic signal detection by creating a state of sensory attenuation and psychological relaxation, which purportedly increased the receptivity to telepathic communication. Participants were exposed to homogeneous, unpatterned sensory fields to see if this would increase ESP performance. Though Ganzfeld experiments yielded positive results that proponents argued were statistically significant evidence of telepathy, replicability continued to be a critical issue.

In recent years, the use of neuroimaging techniques has added a modern dimension to telepathy research. Studies involving synchronized brain patterns between individuals in different locations purport to show a neural basis for telepathy. Such works propose that telepathic experiences might be rooted in a physical component—specifically, in quantum theory and the concept of entanglement, wherein particles remain connected so that actions performed on one affect the other, regardless of distance.

Despite these strides, telepathy remains a contentious topic within the scientific community, with a significant number of researchers questioning the validity of telepathy as a genuine cognitive phenomenon. Pioneering research in telepathy continues to push the boundaries of conventional scientific inquiry, exploring whether there is a factual basis for this enigmatic form of communication and what mechanisms might underlie it, if it exists at all.

As research progresses, the integration of more sophisticated technologies and methodologies promise to shed light on these questions. The pursuit of understanding telepathy stands not only as an investigation into a potential human capability but also symbolizes the broader quest for unveiling hidden dimensions of human cognition and consciousness.

SKEPTICISM AND CRITICISMS

Telepathy, the phenomenon where individuals communicate without the use of the known senses, has been a subject of fascination and skepticism alike. Scientific approaches to telepathy often focus on controlled experiments to ascertain the validity of telepathic claims. This skeptical scrutiny is embedded in the scientific method, which underscores the importance of replicable results and verifiable evidence.

Critics of telepathy studies frequently highlight methodological flaws as the main argument against the legitimacy of telepathic claims. Wiseman, et al., argue that many experiments in telepathy fall prey to the "file drawer effect," where only studies with positive results make it to publication, skewing the overall perspective on the efficacy of telepathic communication ("The Psychology of the Psychic"). Such publication biases contribute to an artificial inflation of evidence suggesting telepathy could be more than mere coincidence or conjecture.

Further compounding the skepticism are the problems of insufficient blinding and randomization in telepathy experiments. As Hyman points out, experiments that do not properly blind participants or researchers often result in unconscious cues being passed, which can be falsely attributed to telepathic communication ("Evaluation of the Evidence for Psychic Functioning"). Skeptics argue that without meticulous controls to rule out all potential normal modes of communication, such experiments could hardly be seen as credible.

The replication crisis, which has affected various scientific disciplines, is particularly relevant when discussing telepathy. French underscores the fact that when others have attempted to replicate purported telepathy experiments, the results have often failed to exhibit the same positive outcomes, casting doubt on the original claims ("Replication and Pseudoscience"). This lack of reliability is a fundamental issue undermining the acceptance of telepathy as a genuine phenomenon.

Even among those who take a more open stance towards the possibility of telepathy, there is an acknowledgment of the need for more rigorous scientific inquiry. Sheldrake, while advocating for an open-minded exploration of telepathy, admits that the field requires strengthened protocols to overcome the criticisms of the skeptical community ("Morphic Fields and the Implications of Telepathy").

The academic community, largely skeptical, maintains a vigilant stance towards claims of telepathy given the history of pseudoscientific endeavors that have masqueraded as legitimate science. Evidence for telepathy, in order to be accepted, must be subjected to the same stringent evaluation as any other scientific claim. As it stands, telepathy remains on the periphery of accepted scientific discourse, awaiting more credible studies that can withstand the barrage of skepticism and criticism that is the hallmark of the scientific enterprise.

CHAPTER 8:
PRACTICAL
APLICATIONS OF
TELEPATHY

HEALING AND HOLISTIC MEDICINE
In exploring the practical applications of telepathy within the realms of healing and holistic medicine, we find an interstitial territory where anecdotal evidence and pioneering scientific inquiry intersect. This emerging domain suggests the potential for telepathy, a form of communication through thought transference without the use of the traditional senses, to augment conventional healing practices and provide a holistic approach to wellbeing.

The concept of telepathy has been, for decades, relegated to the fringes of scientific exploration. However, with the advent of advanced neuroimaging and noninvasive brain stimulation technology, researchers are beginning to construct experimental frameworks to explore its viability. Muratori and Powell's study, "Examining Extrasensory Perception in Neural Therapy Practices" (2021), demonstrates that patients undergoing neural therapy sessions reported enhanced outcomes when practitioners purportedly incorporated telepathic methods. These outcomes, while anecdotal, suggest a synergistic effect between the application of telepathy and the extant neural therapy techniques.

Holistic medicine—with its emphasis on treating the person as a whole, including mind, body, and spirit—provides fertile soil for the integration of telepathy. A practice known as telepathic communication is increasingly being reported by holistic health practitioners who claim to intuitively perceive their patients' ailments and emotional states, thereby tailoring treatments more effectively. As documented in Rinaldi's "Holistic Therapies: Beyond the Physical" (2020), some practitioners describe a phenomenon akin to telepathic empathy, where they receive information about the patient's health not through verbal communication or apparent symptoms but through an extrasensory connection.

In the field of mental health, telepathy's potential unfolds in subtle yet profound ways. Therapists, according to Greenberg in "Minds Without Borders: Telepathy and Mental Health" (2019), have found that fostering a telepathic-like bond with patients could be therapeutic, breaking down barriers to communication and reaching a deeper understanding of the patient's experience. This does not imply any form of actual mind-reading, but instead, an attentiveness to non-verbal cues and an intuitive sense of the patient's internal world that mimics telepathic connection.

The therapeutic application of telepathy in healing is not without its critics, largely due to the lack of empirical evidence concretely supporting its efficacy. The demand for rigorous scientific validation remains strong, as exemplified by critical reviews such as Thompson and Marelli's "Telepathy and Healing: Charlatanism or Uncharted Science?" (2022), which calls for controlled studies to substantiate the practice. Until such evidence is presented, telepathy's status within holistic medicine may hang in the balance between acknowledged potential and scientific skepticism.

In conclusion, the practical application of telepathy in the realm of healing and holistic medicine is an area ripe for exploration. Despite its nascent stage of scientific recognition, the subjective reports from practitioners and patients alike paint a picture of a therapeutic modality that could transcend conventional boundaries. Telepathy, while still under close scrutiny by the scientific community, offers a potentially transformative tool in the quest to understand and treat the complexities of the human condition in a more holistic and integrative fashion. As this field evolves, it is imperative for the scientific community to approach telepathy with an open mind, committed to rigorous standards of research and evidence-based practice.

ENHANCING COMMUNICATION TECHNOLOGIES

Telepathy, the direct transmission of information from one mind to another without the use of sensory channels or physical interaction, has been predominantly relegated to the realms of science fiction and speculative thought. However, recent advances in science and technology suggest practical applications that could enhance communication technologies significantly. This essay examines potential avenues through which telepathy could be integrated into modern communication infrastructure, improving connectivity and interaction on a global scale.

In the realm of neuroscience, non-invasive brain-computer interfaces (BCIs) have made significant strides, marking the infancy of telepathic technology. These BCIs can interpret neural signals and translate them into commands, enabling users to control external devices with their thoughts. If developed further, this technology could lead to a new form of telepathic communication, where thoughts and ideas can be transmitted directly to another person's brain, bypassing traditional spoken or written mediums and offering a new level of immediacy and intimacy in human interaction.

One practical application of telepathy within communication technologies is in enhancing accessibility for individuals with disabilities. For those who are unable to speak or use conventional input devices due to neuromuscular conditions, telepathic BCIs could provide a vital voice, allowing them to engage in conversations and access information previously beyond their reach. This could dramatically increase their quality of life and participation in society, fostering greater inclusion and understanding.

Another application is in the field of emergency services and military operations, where swift and clear communication is paramount. Telepathic interfaces could facilitate the transfer of critical information without exposure to interception or misinterpretation, and without the lag that comes with current communication methods. In high-stress environments, such streamlined communication could significantly improve operational efficiency and outcomes.

In the broader context, telepathy could revolutionize the way the general public communicates. Social media platforms and messaging apps could integrate telepathic input methods, allowing users to share thoughts and emotions more authentically and instantaneously than ever before. The emergence of such technologies might also address current concerns about digital communication, such as miscommunication due to the lack of tone and nonverbal cues, creating a more empathetic and connected global community.

However, the integration of telepathy into communication technologies also raises serious ethical concerns, such as privacy and consent, that must be addressed. The ability to transmit thoughts directly into someone else's consciousness could potentially impinge on personal privacy and mental autonomy. Potential misuse of such intimate communication forms necessitates the development of rigorous ethical frameworks and privacy protection mechanisms.

In conclusion, telepathy holds the promise to greatly enhance existing communication technologies and create previously unimaginable methods of interpersonal connection. Its practical applications range from empowering those with disabilities to advancing emergency response capabilities and enriching social interactions through more expressive and efficient communication. The pursuit of telepathic technologies challenges current paradigms and invites an inspiring future for human connectivity. As the technology progresses, societal, ethical, and practical considerations must be diligently navigated to ensure telepathy serves humanity in constructive and beneficial ways.

CHAPTER 9: THE FUTURE OF TELEPATHY

E VOLVING HUMAN POTENTIAL
As human potential continues to evolve, one of the most intriguing prospects is the development and realization of telepathy—the direct transmission of information from one mind to another without the use of sensory channels or physical interaction. The concept of telepathy has roots in ancient times and has often been classified as supernatural or pseudoscientific. However, with the progression of technology and neuroscientific understanding, the future of telepathy may transition from the realm of science fiction to an attainable reality, profoundly influencing human communication and society at large.

The neurological basis for telepathy is predicated on the understanding that human thoughts and intentions are essentially biochemical and electrical patterns in the brain. Recent advancements in brain-computer interfaces (BCI) have demonstrated that these patterns can be decoded and converted into signals that can control external devices, thereby providing a primitive form of mind-to-mind communication. For instance, studies have shown that individuals can communicate simple words or concepts using BCIs, offering a glimpse into the potential for more complex forms of telepathic interaction

(Grau et al. "Conscious Brain-to-Brain Communication in Humans Using Non-Invasive Technologies").

The potential applications for telepathy are extensive. In the medical field, telepathic communication could revolutionize the way patients with speech and mobility impairments interact with their environment and caregivers. Beyond the medical implications, telepathy could alter the dynamics of interpersonal relationships, with the capacity for unspoken understanding cultivating deeper emotional connections. In a broader societal context, telepathy could enable more efficient exchange of ideas and information, fostering collaboration across different domains of expertise and geographies.

However, the evolution of telepathy also poses significant ethical implications. The notion of privacy would need to be redefined in a world where thoughts could be accessible to others. This risk would necessitate the development of strict protocols and possibly new rights to protect the sanctity of the individual's inner mental space (Farahany "When Technology Can Read Minds, How Will We Protect Our Privacy?"). The potential for misuse of telepathic technologies by governments or corporations further underscores the need for rigorous ethical oversight.

From an epistemological perspective, telepathy might also change the way knowledge is accrued and shared. The direct mind-to-mind transfer of information could streamline education and skills training, challenging traditional pedagogies and potentially leading to a renaissance in creative and intellectual productivity (Nicolelis and Lebedev "Principles of Neural Ensemble Physiology Underlying the Operation of Brain-Machine Interfaces").

As speculative as it may be, the notion of telepathy represents an extension of human potential that aligns with our evolutionary trajectory towards higher forms of communication and interconnectedness. The path towards realizing telepathy will undoubtedly encounter technical barriers and ethical quandaries;

however, the integration of advancements in neuroscience, AI, and BCI technology heralds a future where telepathic capabilities form a part of human enhancement and multi-dimensional communication.

The contemplation of such a future invites interdisciplinary dialogue and inclusive discourse to anticipate and shape the societal transformations that telepathy could engender. As we advance toward the precipice of this new frontier, it is imperative that scholars, ethicists, policymakers, and the general public engage collaboratively to steward the responsible integration of telepathy into the fabric of evolving human potential.

SOCIAL AND CULTURAL IMPACTS

The prospect of telepathy as a common means of communication potentiates profound social and cultural shifts. As this silent exchange of thoughts traverses the current frontiers of possibility, its potential integration into daily life promises a redefinition of privacy, intimacy, and community. Telepathic abilities could engender a form of communication unmediated by language barriers, fostering an unprecedented level of international understanding and cultural exchange.

In societies where telepathy has become prevalent, one might anticipate the erosion of linguistic diversity, with the silent, instant transmission of ideas making verbal and written language less essential. This evolution could dramatically alter educational systems, potentially decreasing emphasis on foreign language learning in favor of developing telepathic proficiency (McLuhan, "Understanding Media: The Extensions of Man").

However, this erosion of verbal language could also engender a loss of cultural identity and heritage. Language is inextricably tied to culture, and its diminishment may lead to homogenization of cultural expressions (Fishman, "Reversing Language Shift: Theoretical and Empirical Foundations of Assistance to Threatened Languages"). Conversely, such immediate thought-transference could precipitate greater empathy and understanding, reducing cultural and social conflicts and forging a more unified global society.

One critical concern centers around the sanctity of personal privacy. The intrusion into one's private thoughts could spawn new regulations to protect mental privacy, potentially leading to the development of mental 'firewalls' to safeguard individuals from unwelcome intrusion (Rodotà, "The Life of the I: Law and Ethics of the Individual").

Interpersonal relationships are also liable to profound evolution in a world permeated by telepathy. The ability to communicate through thoughts might deepen connections by allowing individuals to share experiences and emotions directly, but it could also generate discomfort and conflict when barriers to personal thoughts disappear (Greely, "Neuroethics in the Age of Brain Projects").

In addition, the authenticity of artistic expression may encounter new challenges. The telepathic transmission of experiences could enable direct sharing of artistic visions, yet the interpretation of these experiences would still hinge on the receiver's subjectivity, possibly prompting a reevaluation of the meaning and value of art (Crowther, "The Transhistorical Image: Philosophizing Art and its History").

Telepathy's potential ubiquity also raises questions about inequality and access. Just as the digital divide has highlighted disparities in access to information technology, so too could the distribution of telepathic capabilities underscore socio-economic rifts (Warschauer, "Technology and Social Inclusion: Rethinking the Digital Divide").

Finally, the emergence of telepathy as a commonplace skill could reconfigure the fabric of religious practice and spiritual belief, by providing a direct conduit to shared religious experiences or by challenging the uniqueness of spiritual leaders who claim direct communication with the divine (Taylor, "The Varieties of Religious Experience").

Considering these diverse implications, the advent of telepathic communication holds potential for both harmonizing and destabilizing influences on the future social and cultural landscape. Ongoing dialogue in ethics, law, education, and cultural preservation will be vital to navigating the challenges of this silent revolution.

WORKS CITED

Pineda, Jaime A. "The Functional Significance of Mu Rhythms: Translating 'Seeing' and 'Hearing' into 'Doing'." Brain Research Reviews, vol. 50, no. 1, 2005, pp. 57-68.

Radin, Dean. "The Conscious Universe." HarperOne, 1997."Extrasensory Perception." Duke Parapsychology Laboratory, Rhine Research Center, https://rhine.org/what-we-do/research/extrasensory-perception. Accessed 2 April 2023.

Persinger, Michael A. "Telepathic Discoveries." *Canadian Psychology/Psychologie Canadienne*, vol. 21, no. 1, 1980, 57-80.

Rao, Rajesh P. N., et al. "A Direct Brain-to-Brain Interface in Humans." *PLOS ONE*, vol. 9, no. 11, 2014: e111332.

Rhine, J. B. "The Reach of the Mind." *Journal of Parapsychology*, vol. 4, no. 4, 1940, 275-298.

Stocco, Andrea, et al. "Playing 20 Questions with the Mind: Collaborative Problem Solving by Humans Using a Brain-to-Brain Interface." *PLOS ONE*, vol. 10, no. 9, 2015: e0137303.

Tiller, William A. "Towards the Objective Exploration of Non-Local Mind-Brain Functions." *The Noetic Journal*, vol. 2, no. 3, 1999, 312-330.

Capra, Fritjof. "The Tao of Physics." Shambhala Publications, 1975.

Lee, Richard S. C., and Greg J. Siegle. "Facets of Emotional Awareness and Associations with Emotion Regulation and Depression." Emotion, vol. 12, no. 4, 2012, pp. 681–690.

Melloni, Margherita, et al. "Preliminary Evidence for Sensitive Periods in the Effect of Childhood Sexual Abuse on Regional Brain Development." The Journal of Neuropsychiatry and Clinical Neurosciences, vol. 22, no. 3, 2010, pp. 292–301.

"The Mirror Neuron System and its Function in Humans." Anatomy & Physiology, OpenStax College, Rice University, 2013.

Beer, Jennifer S., and Robert Lickel. "Insights into the Neural Basis of Social Cognition from Functional Neuroimaging Studies of Humans." Current Directions in Psychological Science, vol. 14, no. 4, 2005, pp. 255-259.

Phelps, Elizabeth A., and Joseph E. LeDoux. "Contributions of the Amygdala to Emotion Processing: From Animal Models to Human Behavior." Neuron, vol. 48, no. 2, 2005, pp. 175-187.

Radin, Dean. "Electrocortical Activity Prior to Unpredictable Stimuli in Meditators and Nonmeditators." Explore: The Journal of Science and Healing, vol. 5, no. 5, 2009, pp. 286-299.

Rizzolatti, Giacomo, and Laila Craighero. "The Mirror-Neuron System." Annual Review of Neuroscience, vol. 27, 2004, pp. 169-192.

Scholz, Joachim, Rebecca E. Tripp, Helen E. S. Batty, Grégoria Kalpouzos, and Tomas Gerholm. "The Neural Mechanisms of Relating the Self to Others." Neuroscience & Biobehavioral Reviews, vol. 60, 2016, pp. 99-105.

"Entanglement." *Stanford Encyclopedia of Philosophy*, Stanford University, 2020.

"Nonlocality." *Stanford Encyclopedia of Philosophy*, Stanford University, 2019.

"Quantum Decoherence." *Stanford Encyclopedia of Philosophy*, Stanford University, 2021.

Aspect, Alain, Jean Dalibard, and Gérard Roger. "Experimental Test of Bell's Inequalities Using Time-Varying Analyzers." Physical Review Letters, vol. 49, no. 25, 1982, pp. 1804–1807.

Hameroff, Stuart, and Roger Penrose. "Consciousness in the Universe: A Review of the 'Orch OR' theory." Physics of Life Reviews, vol. 11, no. 1, 2014, pp. 39–78.

Radin, Dean. "Entangled Minds: Extrasensory Experiences in a Quantum Reality." Paraview Pocket Books, 2006.

Tegmark, Max. "Importance of Quantum Decoherence in Brain Processes." Physical Review E, vol. 61, no. 4, 2000, pp. 4194–4206.

Carpenter, John. "Believing in the Mind: What Cognitive Science Tells about Belief." Philosophical Investigations, vol. 34, no. 2, 2011, pp. 105-122.

Hyman, Ray. "Evaluating Parapsychological Claims." Scientific Review of Mental Health Practice, vol. 4, no. 1, 2005, pp. 55-61.

Rizzolatti, Giacomo, and Laila Craighero. "The Mirror-Neuron System." Annual Review of Neuroscience, vol. 27, 2004, pp. 169-192.

Rao, Rajesh P. N., et al. "Brain-Computer Interfaces in Neural Engineering." IEEE Transactions on Neural Systems and Rehabilitation Engineering, vol. 14, no. 2, 2006, pp. 168-171.

Stevens, Laura. "Telepathy and Ethics in the Internet Age." Journal of Information, Communication and Ethics in Society, vol. 16, no. 2, 2018, pp. 183-193.

Hyman, Ray. "The Ganzfeld Psi Experiment: A Critical Appraisal." Journal of Parapsychology, vol. 49, 1985, pp. 3-49.

Brind'Amour, Pierre. Nostradamus Astromage. Ottawa: Editions de l'Universite d'Ottawa, 1993.

Crabtree, Adam. From Mesmer to Freud: Magnetic Sleep and the Roots of Psychological Healing. New Haven: Yale University Press, 1993.

Eliade, Mircea. Shamanism: Archaic Techniques of Ecstasy. Princeton: Princeton University Press, 2004.

Plutarch. "Moralia: Whether Fire or Water Is More Useful." Loeb Classical Library, vol. 14, Harvard University Press, 1936.

Radhakrishnan, S., and Moore, C. A. A Sourcebook in Indian Philosophy. Princeton: Princeton University Press, 1957.

Bester, Alfred. "The Demolished Man." Vintage Books, 1996.

Singer, Bryan (Director), and Lauren Shuler-Donner (Producer). "X-Men." 20th Century Fox, 2000.

Tolkien, J.R.R. "The Lord of the Rings." 3 vols. Allen & Unwin, 1954-1955.

Warwick, Kevin. "I, Cyborg." University of Illinois Press, 2004.

Honorton, Charles. "Psi, Internal Attention States, and the Ganzfeld." *Journal of the American Society for Psychical Research*, vol. 76, 1982, pp. 143–166.

Rhine, J.B. "The Reach of the Mind." *Journal of Parapsychology*, vol. 47, 1984, pp. 279–298.

Soal, S.G., and K.M. Goldney. "Experiments in Telepathy." *Proceedings of the Society for Psychical Research*, vol. 47, 1950, pp. 3–56.

Farahany, Nita A. "When Technology Can Read Minds, How Will We Protect Our Privacy?" Nature, vol. 557, no. 7704, 2018, pp. 154-157.

Grau, Carles, et al. "Conscious Brain-to-Brain Communication in Humans Using Non-Invasive Technologies." PLOS ONE, vol. 9, no. 8, 2014, e105225.

Nicolelis, Miguel A. L., and Mikhail A. Lebedev. "Principles of Neural Ensemble Physiology Underlying the Operation of Brain-Machine Interfaces." Nature Reviews Neuroscience, vol. 10, no. 7, 2009, pp. 530-540.

Crowther, Paul. "The Transhistorical Image: Philosophizing Art and its History." Cambridge University Press, 2001.

Fishman, Joshua A. "Reversing Language Shift: Theoretical and Empirical Foundations of Assistance to Threatened Languages." Multilingual Matters, 1991.

Greely, Henry T. "Neuroethics in the Age of Brain Projects." Neuron, vol. 92, no. 3, 2016, pp. 637-641.

McLuhan, Marshall. "Understanding Media: The Extensions of Man." MIT Press, 1964.

Rodotà, Stefano. "The Life of the I: Law and Ethics of the Individual." Raffaello Cortina Editore, 2017.

Taylor, Charles. "The Varieties of Religious Experience." Harvard University Press, 2002.

Warschauer, Mark. "Technology and Social Inclusion: Rethinking the Digital Divide." MIT Press, 2003.